Radiation protection off-site for emergency services in the event of a nuclear accident

Contents

Introduction *1*

Responsibilities *1*

Dose limitation *1*

Dose assessment and control *2*

Protective measures *2*
Potassium iodate tablets *2*
Respiratory protective equipment *2*
Protective clothing *3*
Control of body surface contamination *3*

Monitoring and recording of doses *3*
Monitoring of doses *3*
Recording of doses *3*

Training *4*

Appendix 1 The off-site radiological hazard *5*

Appendix 2 References *6*

Appendix 3 Explanation of terms *6*

London: HMSO

©*Crown copyright 1991*
First published 1991

HS(G) series

Further advice on this or any other HSE publications may be obtained from Area Offices of HSE or from the general enquiry points listed below.

Health and Safety Executive
Library and Information Services
Broad Lane
SHEFFIELD S3 7HQ
Telephone: (0742) 75253
Telex: 54556

Health and Safety Executive
Library and Information Services
Baynards House
1 Chestow Place
Westbourne Grove
LONDON W2 4TF
Telephone: 071-221 0870
Telex: 25683

ISBN 0 11 885576 X

Introduction

1   This booklet provides guidance on the control of radiation exposure to emergency services personnel when assisting the public after a serious accident at a civil nuclear installation (hereafter referred to as 'a nuclear site'). The advice applies to both the management and staff of emergency services but only in relation to their work off-site. It takes into account the hazards from radiation that it is reasonable to anticipate and to plan for outside the nuclear site boundary. It is therefore not directly relevant to those nuclear sites where it has been judged that no off-site plan is required.

2   The advice covers situations that might arise after a serious accident at a nuclear site. It will therefore help emergency and essential services to respond positively when they are consulted during the preparation of site specific emergency plans, and should also help them develop their own plans for a response to other types of radiation accidents.

3   A framework of principles and guidance is set out, for use as appropriate, by those responsible for emergency planning. It is not possible to give detailed guidance on all the possible variations in accident scenarios that could occur, so modifications may be needed to meet local needs or circumstances (eg geographical or industrial considerations).

4   A Home Office Joint Working Group on Chernobyl has prepared specific advice[1] for firefighters (both on and off-site). This advice has been accepted by the Central Fire Brigades Advisory Councils and should be read in conjunction with this guidance.

5   The HSE booklet *Arrangements for responding to nuclear emergencies*[2] explains, among other things, how emergency plans would be put into effect. It describes the organisational aspects of a planned response to an accident and explains the roles of the different organisations involved.

6   Appendix 1 contains some further information about the likely off-site radiological hazard and risk in the event of a serious accident at a nuclear site.

Responsibilities

7   The operator of a nuclear site carries the responsibility for the safe running of the plant as well as liability for injuries and damage resulting from incidents involving radioactive substances or radiation emitted from the site.

8   However, under the statutory provisions of the Health and Safety at Work etc Act 1974 (HSW Act) and the Ionising Radiations Regulations 1985 (IRR 85), the safety of emergency service personnel undertaking work during an emergency or accident is primarily the responsibility of their employer or controller.

9   Employers of emergency service personnel should ensure that their own plans, taken together with the emergency plans made by site operators, adequately protect the overall health and safety of their staff. The radiation protection standards relevant to workers should be applied to the police, notwithstanding their special employment status. Emergency services should look to the site operator for cooperation and help over the detailed arrangements in plans that concern their protection, particularly in those aspects where the services do not possess the special knowledge, expertise or equipment needed in dealing with such accidents.

10   Those responsible for making emergency plans within the site operator's organisation and the emergency services should agree the role envisaged for the emergency services in the event of a serious accident. Those emergency services who lack in-house expertise in radiation protection may, if they so wish, obtain independent advice from the National Radiological Protection Board (NRPB) or from other radiation protection consultants.

11   When putting an emergency plan into effect, all those involved should coordinate their efforts to ensure their own radiation protection. Emergency services can expect to receive advice from the radiation protection specialists in the site emergency control centre (SECC)*, in the early stages of an emergency, as well as from any expertise they may themselves provide. All emergency plans made by site operators which contain contingencies for an off-site response include provision for a properly equipped off-site centre (OSC)* at a previously organised location. Once the OSC has been established, it will take over the responsibility of passing radiological information and advice to those engaged in the off-site response.

Dose limitation

12   A fundamental radiological protection principle and legal requirement that all exposures to radiation should be as low as reasonably practicable remains valid in an accident situation and should be applied as well as adherence to dose limits.

13   The annual whole body dose limit in IRR 85 for adult employees and people at work is 50 mSv and for planning purposes this figure should be used as the single exposure limit for emergency services personnel

---

* Denotes an entry in the *Explanation of terms* on page 6

working off-site. While dose limits are primarily intended to protect people who are routinely exposed to radiation, current NRPB advice is that doses to such people should not exceed 15 mSv per year averaged over a period of years. The 50 mSv limit for a single emergency exposure for off-site working is thus not inappropriate when planning for very improbable events.

14   Under IRR 85 there are additional dose limits which should be applied for planning purposes for single emergency situations as below:

(a)   for individual organs - 500 mSv;

(b)   for women of reproductive capacity - 13 mSv external dose to the abdomen;

(c)   for pregnant women - 10 mSv external dose to the abdomen. In addition, it is particularly relevant in the case of pregnant women, that intakes of radioactive substances should be avoided if at all possible;

The dose limits mentioned in paragraphs 13 and 14 are not related to emergency reference levels* of dose for the public.

Dose assessment and control

15   When an emergency plan is put into action because a nuclear emergency has been declared, survey teams in suitably equipped vehicles will take the necessary measurements off-site and in a predetermined manner within the detailed emergency planning zone.* In the early stages the site radiation protection team in the SECC will assess the results; in later stages this task will pass to the OSC, and survey teams from other sites, government departments and agencies should also be contributing measurements. At every stage, the liaison officer(s) from the emergency services in the SECC or OSC should be given the necessary advice on radiological protection to pass on to the people who may need to know or act on it.

16   Typical measures that could be advised for the protection of emergency service personnel off-site, depending on the circumstances at the time, include the following:

(a)   *limiting the time spent in the affected areas** and, if necessary, adjusting the staff deployment by eg shift changes to continue essential work. Any such areas would be identified at the time and would be specific locations demarcated by the police cordons or checkpoints;

(b)   *providing direct reading dosemeters* with instructions on the dose at which the person should withdraw (see paragraph 25);

(c)   *making use of protective measures* These include taking potassium iodate tablets for protection against releases to the atmosphere which contain radioactive iodine. Exceptionally, respiratory protective equipment may be needed against radioactive particulate matter (see paragraph 20);

(d)   in rare circumstances *providing a trained health physics adviser to accompany staff* and give immediate advice and radiological supervision. This degree of supervision is not likely to be needed as much off-site as on-site.

Protective measures

Potassium iodate tablets*

17   Where the main radiological risk would come from inhaling airborne radioactive iodine, the most effective countermeasure is to take potassium iodate tablets. These 'block' or reduce the accumulation of radioactive iodine in the thyroid gland. These tablets should, if possible, be taken *before* exposure. Taken in advance, the tablets are almost completely effective in preventing any dose to the thyroid. If exposure occurs first, the tablets will still substantially reduce the thyroid dose if taken as soon as possible afterwards; even six hours after exposure a 50% reduction in potential dose may be achieved. Potassium iodate tablets have no significant side effects for normal healthy adults (but see Appendix 3).

18   For sites where radioactive iodine could present a significant hazard off-site, sufficient stocks of potassium iodate tablets should be held for distribution to the emergency services personnel. These stocks should be kept at agreed distribution points that are identified within the emergency plans and should be renewed before their shelf-life expires.

19   When an emergency has been declared involving radioactive iodine, the appropriate number of tablets should be issued to all personnel likely to enter affected areas. Advice on when to take them will be given via the liaison officer(s) from the emergency services in the SECC or OSC.

Respiratory protective equipment

20   There are many types of respirator, ranging from simple filtering facepieces to full face masks with filters and/or absorbent canisters. Respirators can reduce the amount of radioactive material that such personnel might otherwise inhale, by filtering or treating the air. However, in the case of reactor accidents and in view of the protection afforded by potassium iodate tablets, it is unlikely that respiratory protection would be necessary for off-site personnel. Local circumstances such as a particular type of plant with specific hazards, might justify

the pre-issue of respiratory equipment to local emergency service personnel. Otherwise, sufficient stocks of suitable equipment will be held on site for issue to emergency services if the survey measurements show that this is necessary.

Protective clothing

21   There is no need to plan for the wearing of special protective clothing since normal uniforms will provide adequate protection for the low levels of contamination that can be foreseen off-site; subsequent decontamination will, however, be easier if a simple normal overall or oversuit (or suitable weatherproof clothing in inclement weather conditions) is worn. Emergency services may have access to suitable clothing; if not, they should make prior arrangements with the site operator for provision of such clothing. However, account should also be taken of the Hinkley Point Inquiry Inspector's Report[3] (paragraph 50.129 and recommendation 7B(7)) on the public reassurance provided by police officers calmly performing their duty wearing ordinary police uniform.

Control of body surface contamination

22   On leaving the affected area, personnel should return to a pre-arranged place to remove their outer clothing and place it in a plastic bag; the site operator will arrange for decontamination or disposal of such clothing as appropriate. Shoes should be wiped clean and the used tissues or cloths similarly bagged. The person should then wash thoroughly, at or before going, to the nearest reception centre or other centre at which the Health Authority has made provision for personnel monitoring, to check that no significant contamination remains; acceptable levels (instrument readings) of remaining contamination will depend on the radionuclides released in the accident and the calibration of the instruments used, therefore numerical values are not specified in this guidance. The health physics staff at the centre and the site health physicists will have the necessary expertise in this matter and will advise if any further action is required.

Monitoring and recording of doses

Monitoring of doses

23   Arrangements should be made with site operators for the prior delivery of sufficient stocks of dosemeters (normally of the film badge or TLD type) from an approved dosimetry service to one or more centres, for subsequent distribution by each emergency service to those of its personnel who will be working in the affected area. However, some personnel may be deployed direct to their duties and in these situations they should be provided with dosemeters as soon as practicable; doses will be calculated subsequently (see paragraph 27). Emergency services should keep a record of each dosemeter, to whom it was issued, in what locations and for how long it was worn (see paragraph 28). Dosemeters should be collected before entering and handed in on leaving the affected area following the control procedures set up by each relevant emergency service in line with their pre-planning. Pre-planning should include provision for the issue of guidance as to any need for the supply of fresh dosemeters in exchange for dosemeters which need to be recalled for assessment.

24   The dosemeters should normally be worn on the trunk of the body, inside the outermost layer of clothing and throughout the duration of the emergency unless the liaison officer(s) from the emergency services passes on advice that it is no longer necessary to wear them.

25   When police officers need to remain behind at inner cordon positions (checkpoints) for long periods they should, in addition to the above, be issued with a direct reading or alarm dosemeter to let them check external exposure. This monitoring is intended to give reassurance beyond that provided by the liaison officer and it is acceptable for site operators to arrange for these dosemeters to be available for issue as soon as resources permit.

26   Where radioactive iodine may have been released, emergency services personnel, before going off duty, should undergo thyroid monitoring at the reception centre *either* if such action is officially advised *or* the officer personally wishes it. This will establish the level of exposure due to the intake of radioactive iodine (see paragraph 17), and is therefore only appropriate to such incidents.

27   In addition to the assessments of dose indicated in paragraphs 23, 24 and 26, measurements from the continuous surveys undertaken in the area can be used, together with the information of where and for how long an officer has spent in a particular location, to assess radiation dose. In special cases, additional personal dosimetry techniques can also be used to provide information about internal dose.

Recording of doses

28   Individual emergency services should each designate an officer to be responsible for keeping a record of relevant information of personnel working in the affected areas for control purposes and for future reference as necessary. Records should show the time spent in the affected areas, locations, details of personal protective measures employed (eg tablets taken), dosemeters and overalls issued, decontamination efforts and results, and the results of any other monitoring.

29 In the event of a prolonged emergency, arrangements will have to be made to provide dose information from all the various measurements in sufficient time to assess whether officers can return to the affected area for a second or subsequent tour of duty.

30 The relevant approved dosimetry service(s) will subsequently supply copies of dose records, which should be retained on the officer's personal file for 50 years, together with any further dose information obtained from the measurements mentioned in paragraph 27.

Training

31 Emergency services should make arrangements, in conjunction with the site operator, for the adequate training of all staff who would be expected to undertake duties off-site in the emergency zone. The scope of the training should be commensurate with the planned response but would be expected to include:

(a) familiarisation with the general nature of the potential hazards of the particular site;

(b) use and practice of the measures and systems for radiological protection and control; and

(c) practical work with any relevant equipment.

32 Those parts of emergency services that would be expected to be deployed soon after an initial notification of an incident should participate in site emergency exercises on a regular basis.

33 Training requirements should be identified, carried out and subsequently recorded in the officer's personnel records. Any special training needed, ie beyond that normally undertaken by the service, would be best provided by the site operator in conjunction with the emergency service.

Appendix 1 The off-site radiological hazard in the event of a serious accident at a nuclear site

1   In complying with conditions attached to a nuclear site licence the site operator will make arrangements for responding to nuclear emergencies. These require the operator to consult with the local authority and the emergency services. A list of nuclear licensed sites may be obtained from the Department of Energy Library, 1 Palace Street, London SW1E 5HE. IRR 85 impose the requirement to have a contingency plan on non-licensed sites. The emergency plans in which these arrangements are described make reference to the roles or tasks of the emergency and essential services staff such as the evacuation of the public. The nature of any radiological hazard which might be encountered by emergency services in carrying out their tasks would depend on:

(a)   whether a release of radioactivity had in fact occurred (because the declaration of an emergency would not necessarily mean that an off-site release had taken place);

(b)   whether any release was still in progress or had ceased;

(c)   the magnitude and composition of any release; and

(d)   if airborne, its elevation and the weather conditions affecting its dispersion and deposition on the ground.

2   Exposure to radioactive material released in an accident may occur in a number of different ways:

(a)   *direct radiation from radioactive material*   Such material could be present as an airborne cloud, as a deposit on the ground or other surfaces, or as a deposit on the skin or clothing. Direct radiation results in *external* exposure, which can be measured by dosemeters worn on the body;

(b)   *inhalation of radioactive material*   This results in *internal* exposure. Measurement of internal exposure always requires special techniques;

(c)   *ingestion of radioactive material* after it has contaminated foodstuffs or drinking water. This again results in *internal* exposure.

All these potential exposure routes are considered when providing radiological protection to those working in the off-site area and arranging for assessment of their exposure. The relative importance of each potential exposure route will vary according to the radionuclides released and the approach to dose control and radiological safety will reflect this.

3   There is no reasonably foreseeable possibility of doses off-site ever reaching levels that could cause short-term radiation effects, such as immediate skin burns, radiation sickness etc. For the purposes of radiological protection, however, the assumption is made that any radiation dose carries with it some risk. Statistically, there is about a few thousandths of a per cent risk per millisievert of radiation dose of developing a fatal cancer in the long-term (many years). To put this into context, it should be remembered that everyone is exposed continuously to radiation from many natural sources. In the United Kingdom, the average annual dose from *natural* radiation (from the sky, the earth, etc) is about 2.2 millisieverts, whereas that from *all* sources is about 2.5 millisieverts (of which medical exposures account for approximately 0.3 millisieverts). Discharges from nuclear sites give rise to less than 0.001 millisieverts.

4   The possible nature and scale of releases in an emergency is the subject of detailed assessment as part of the formal licensing process or under the requirements of IRR 85 for every nuclear site. Some common basic features of the various scenarios have been identified:

(a)   the declaration of an emergency at a nuclear site does not automatically signify an immediate or impending serious off-site risk. The most likely situation would be one where the off-site risk was either non-existent or so minor that it would not justify taking any countermeasures to protect the public (ie neither sheltering indoors, nor taking potassium iodate tablets nor evacuating the area would be necessary). The design standards that nuclear installations must meet are such that serious accidents (ie those which could lead to the need to protect the public) are extremely unlikely;

(b)   nevertheless, occasionally, immediate short-term countermeasures such as evacuation may have to be considered. The area then affected would be determined by the magnitude of the release and local conditions such as weather. Generally people upwind of the release point would not be at risk. Detailed emergency planning zones vary in size but typically extend over a few kilometres and the emergency planning is such that a response can be extended over a larger area if necessary;

(c)   if people are exposed to airborne radioactive material there may be an inhalation hazard as well as a direct radiation hazard. After the release has stopped, any inhalation hazard will be much less important, while the main hazard (apart from ingestion via foodstuffs) is likely to be direct radiation from any radioactive materials that have settled on the ground or other surfaces. Those working in the affected area should be instructed not to consume food or drink which may have been exposed to possible contamination, thereby avoiding any ingestion of radioactive substances;

(d) the emergency plans which have been drawn up by site operators recognise these features of accidents and contain off-site hazard action levels which are set to allow time for countermeasures to be advised and undertaken in accordance with NRPB advice on emergency reference levels.

Appendix 2   References

1   Report of the Joint Working Group on Chernobyl, Home Office, 1989.

2   HSE Booklet *Arrangements for responding to nuclear emergencies* - HMSO, ISBN 0 11 885525 5.

3   *The Hinkley Point Public Inquiries; A Report by Michael Barnes QC* HMSO, ISBN 0 11 412955 X.

Appendix 3   Explanation of terms

Affected area   This is the area within the inner police cordon. It is a control zone which has been instituted as a result of the accident but its establishment does not necessarily imply a radiological hazard within it.

Detailed emergency planning zone   This is a defined zone closely surrounding each installation within which arrangements to protect the public should be planned in detail. The boundary of this zone is defined in relation to the maximum size of any accident which can reasonably be foreseen.

Emergency reference levels (ERLs)   Levels of radiation dose specified by NRPB for the introduction of countermeasures to protect the public, published by NRPB as Documents of the NRPB, Volume 1, No.4. They are specified in terms of the likely dose to an individual that could be averted if the countermeasure is taken.

Off-site centre (OSC)   This is an ancillary emergency facility located at some distance from the nuclear site that can be staffed and made operational within a few hours of an emergency being declared. Representatives of all those organisations who are involved in the off-site response attend the OSC, including the operator, emergency services, local authority and relevant central government departments and agencies. Prior to the OSC becoming operational, technical advice on off-site aspects of the emergency would be provided from the SECC (see below).

Potassium iodate tablets   These are tablets containing chemically bound iodine in a stable form which permits extended storage. The normal adult dose of 170 milligrams may cause minor effects (headaches and cold-like symptoms) in some individuals and should not be administered to those individuals known to be sensitive to iodine or to those with a history of thyroid disease.

Site emergency control centre (SECC)   This is where all the activities associated with the on-site emergency response are coordinated. It is a specially equipped facility on the site and is provided with maps, charts, telecommunications and all other necessary equipment.

Printed in the UK for HMSO

C50   4/91